I0483157

DISCLAIMER

Mention of any company or product does not constitute endorsement by the National Institute for Occupational Safety and Health (NIOSH). In addition, citations to Web sites external to NIOSH do not constitute NIOSH endorsement of the sponsoring organizations or their programs or products. Furthermore, NIOSH is not responsible for the content of these Web sites.

ORDERING INFORMATION

This document is in the public domain and may be freely copied or reprinted. To receive NIOSH documents or more information about occupational safety and health topics, contact NIOSH at

1–800–CDC–INFO (1–800–232–4636)
TTY: 1–888–232–6348
E-mail: cdcinfo@cdc.gov

or visit the NIOSH Web site at **www.cdc.gov/niosh**

For a monthly update on news at NIOSH, subscribe to
NIOSH eNews by visiting **www.cdc.gov/niosh/eNews**

DHHS (NIOSH) Publication Number 2008–128 (supersedes 2008–113)

May 2008

SAFER • HEALTHIER • PEOPLE™

Protecting Poultry Workers from Avian Influenza (Bird Flu)

WARNING!

During an outbreak of avian influenza (bird flu), poultry workers* may become seriously ill or die after contact with infected poultry or contaminated materials.

All poultry workers and all owners and operators of poultry operations should take the following steps to protect themselves from avian influenza.

Take the following steps BEFORE an outbreak of avian influenza:

1. Make sure that an avian influenza response plan has been developed to complement regional, State, and industry plans.

    ■ Use the CDC guidance presented in the full Alert to develop a response plan.

    ■ Select a response plan manager.

    ■ Coordinate your avian influenza response plan with appropriate State animal and public health agencies.

    ■ Make sure that workers are aware of the avian influenza response plan and their responsibilities.

2. Follow biosecurity practices to keep avian influenza and other diseases out of your poultry flock:

■ Keep your poultry flock isolated from outside environments.

■ Prevent flocks from contacting wild birds and water sources that might be contaminated by wild birds.

■ Allow only essential workers and vehicles to enter the farm and poultry houses.

■ Provide clean protective clothing, equipment, and disinfection facilities for workers.

■ Thoroughly clean and disinfect equipment and vehicles entering and leaving the farm. Include tires and undercarriage.

■ Do not borrow equipment or vehicles from other farms and do not lend yours.

■ Avoid visiting other poultry farms.

■ If you do visit another farm or a live-bird market, change footwear and clothing before contacting your own flock again.

---

*Poultry workers include all workers who may contact poultry or materials or environments contaminated by poultry.

- Do not bring birds from slaughter channels (especially live-bird markets) back to the farm.

3. Know the signs of avian influenza in poultry.

   - Be aware of the signs of avian influenza infection with the highly pathogenic H5N1 virus in poultry:

     — Sudden death without clinical signs or symptoms

     — Lack of coordination

     — Lack of energy and appetite

     — Soft-shelled or misshapen eggs

     — Decreased egg production

     — Purple discoloration of the wattles, combs, and legs

     — Swelling of the head, eyelids, combs, wattles, and hocks

     — Diarrhea

     — Nasal discharge

     — Coughing and sneezing

4. Report sick or dying birds immediately!

   - Immediately report any sick or dying birds in your poultry flock to the proper authorities:

     — Call the U.S. Department of Agriculture (USDA) toll free at 1–866–536–7593.

     — Or contact your State veterinarian or local extension agent.

   - Don't wait to report sick birds! Early detection of avian influenza is essential to prevent its spread.

5. Know the possible signs and symptoms of avian influenza in humans infected with the highly pathogenic H5N1 virus:

   - Fever

   - Cough

   - Shortness of breath

   - Sore throat

   - Muscle aches

   - Conjunctivitis (redness, swelling, and pain in the eyes and eyelids)

   - Diarrhea

6. Consider getting the current season's flu shot.

7. Train workers in all guidance and recommendations presented in this Alert.

Take the following steps DURING an outbreak of avian influenza:

1. Follow the avian influenza response plan.

2. Ask your doctor about taking antiviral medication.

3. Wear personal protective clothing.

   - Wear the following personal protective clothing if you may be exposed to an avian influenza virus:

     — Outer garments (aprons or coveralls)

     — Gloves

(Over)

- Foot protection (boots or boot covers)

- Head protection (head cover or hair cover)

- Choose disposable, impermeable, lightweight protective clothing.

- Wear disposable, lightweight, waterproof gloves (8- to 12-mil nitrile or vinyl, for example) or heavy-duty, 18-mil rubber gloves that are reusable after disinfection.

- Use disposable boot covers or boots that can be reused after disinfection.

- Use disposable head covers or hair covers.

4. Wear eye protection.

- When working with poultry, wear unvented or indirectly vented safety goggles, a respirator with a full facepiece, or a powered, air-purifying respirator (PAPR) with a loose-fitting hood or helmet and face shield.

- Remove eye protection carefully to prevent contaminated equipment from contacting eyes, nose, or mouth.

5. Wear respirators.

- Wear a NIOSH-approved, air-purifying respirator with a particulate filter whenever you are working in poultry barns or may be exposed to infected poultry or virus-contaminated materials or environments.

- Use a particulate filter that is rated N–95 or better.

6. Follow a written respiratory protection program. To make sure that respirators protect workers adequately from avian influenza, do the following:

- Designate a qualified person to oversee the program and answer workers' questions.

- Provide workers with respirator training and fit-testing to assure a safe and comfortable seal for the respirator facepiece.

- Include all workers who may be at risk of exposure to avian influenza virus.

7. Protect yourself when removing personal protective clothing or equipment.

- With your respirator, goggles, and gloves on, remove all personal protective clothing.

- Place disposable clothing in a secure container† for disposal.

- Place reusable clothing in labeled or color-coded bags or containers‡ for cleaning and disinfection.

- Remove gloves and discard them in a secure container.

- Thoroughly wash hands with soap and water.

- If no hand-washing facilities are available, use waterless soaps or alcohol-based sanitizers provided by your employer.

---

†NIOSH recommends the use of secure containers such as those described in the *Code of Federal Regulations* [29 CFR 1910.1030(d)(4)(iii)(B)]. Such containers should therefore be (1) closable, (2) constructed to contain all contents and prevent leakage of fluids during handling, storage, transport, or shipping, (3) labeled or color-coded, and (4) closed before removal.

‡NIOSH recommends the use of labeled or color-coded bags such as those described in the *Code of Federal Regulations* [29 CFR 1910.1030(d)(4)(iv)(A)(2)].

- Be careful about using waterless soaps or alcohol-based sanitizers too often, as they can be very harsh to the skin.

■ Next, carefully remove your goggles and then your respirator.

- Thoroughly wash hands again with soap and water.

- If no hand-washing facilities are available, use waterless soaps or alcohol-based sanitizers provided by your employer.

8. Use good hand hygiene (proper use of gloves, hand-washing, and waterless hand sanitizers) and the decontamination procedures outlined here to prevent infection, avoid taking viruses home, and keep them from spreading to other farms.

9. Shower at the end of the work shift and leave all contaminated clothing and equipment at work.

■ Shower at the worksite or at a nearby decontamination station.

■ Never wear contaminated clothing or equipment outside the work area.

10. Participate in health surveillance and monitoring programs.

■ Make sure a surveillance program has been established to identify symptomatic workers for 10 days after exposure to infected birds or virus-contaminated materials or environments.

■ Seek immediate medical care for workers who develop any of the following symptoms within 10 days of exposure to infected or exposed birds or to virus-contaminated materials or environments:

- Fever

- Cough

- Shortness of breath

- Sore throat

- Muscle aches

- Conjunctivitis (eye infections)

- Diarrhea

■ Tell the health care provider about the possible avian influenza exposure before the ill person arrives.

■ Promptly report suspected human cases to supervisors and to local health authorities.

For additional information, see **NIOSH Alert: Protecting Poultry Workers from Avian Influenza (Bird Flu)** [DHHS (NIOSH) Publication No. 2008–128 (supersedes 2008–113)]. To request single copies of the Alert, contact NIOSH at

1–800–CDC–INFO (1–800–232–4636)
TTY: 1–888–232–6348
E-mail: cdcinfo@cdc.gov

or visit the NIOSH Web site at
www.cdc.gov/niosh

For a monthly update on news at NIOSH, subscribe to *NIOSH eNews* by visiting **www.cdc.gov/niosh/eNews**

DEPARTMENT OF HEALTH AND HUMAN SERVICES
Centers for Disease Control and Prevention
National Institute for Occupational Safety and Health

Proteja de la gripe aviar a los trabajadores de las granjas avícolas

Todos los trabajadores avícolas y los propietarios y operadores de procesos avícolas deben tomar las siguientes medidas para protegerse de la gripe aviar.

Tome las siguientes medidas ANTES de un brote de gripe aviar:

1. Asegúrese de que se ha elaborado un plan para enfrentar la gripe aviar como complemento a los planes regionales, estatales, e industriales.

   - Para elaborar este plan de respuesta, utilice las instrucciones presentadas por los CDC en el documento completo de la Alerta.

   - Designe a un encargado del plan de respuesta.

   - Coordine su plan de respuesta ante la gripe aviar con las correspondientes agencias estatales de control animal y salud pública.

   - Asegúrese de que los trabajadores estén al tanto de este plan y de sus responsabilidades.

2. Siga las prácticas de bioseguridad que se presentan a continuación para mantener a su plantel avícola libre de gripe aviar y de otras enfermedades:

   - Mantenga a su plantel avícola aislado de los ambientes externos.

   - Evite el contacto de las aves de corral con las aves silvestres y con fuentes de agua que pueden haber sido contaminadas por las aves silvestres.

   - Permita que solamente entren a la granja y a las jaulas el personal y los vehículos esenciales.

   - Proporcione a los trabajadores ropa protectora limpia, equipo e instalaciones para la desinfección.

   - Limpie y desinfecte a fondo los equipos y los vehículos (incluidos neumáticos y debajo de los vehículos) que entran y salen de la granja.

   - No utilice equipo ni vehículos de otras granjas, y no preste los suyos.

   - Evite visitar otras granjas avícolas.

   - Si usted visita otra granja o un mercado de aves vivas, cámbiese de ropa y de calzado antes de volver a acercarse a su plantel avícola.

   - No lleve a la granja las aves de los canales de matanza, en especial de los mercados de aves vivas.

---

*Los trabajadores avícolas son todos aquellos que pueden entrar en contacto con aves de corral, materiales, o ambientes contaminados por éstas.

3. Conozca los signos de la gripe aviar en las aves de corral.

   ▣ Manténgase atento a los signos de la infección por gripe aviar en las aves de corral causada por el virus altamente patógeno H5N1:

      — Muerte repentina sin signos ni síntomas clínicos

      — Falta de coordinación

      — Falta de energía y apetito

      — Producción de huevos con cáscaras blandas o deformes

      — Disminución en la producción de huevos

      — Manchas moradas en la barbilla, la cresta, o las patas

      — Inflamación de la cabeza, párpados, cresta, barbilla, o patas

      — Diarrea

      — Secreción nasal

      — Tos y estornudos

4. ¡Reporte inmediatamente los casos de aves enfermas o moribundas!

   ▣ Reporte inmediatamente todo caso de aves enfermas o que se estén muriendo en su plantel avícola a las correspondientes autoridades:

      — Llame al Departamento de Agricultura de EE.UU. (USDA, por sus siglas en inglés) a través de su línea gratuita 1–866–536–7593.

      — O comuníquese con su veterinario estatal o el agente local del Servicio de Extensión Agrícola.

   ▣ ¡No espere para reportar las aves enfermas! La detección temprana de la gripe aviar es esencial para prevenir su transmisión.

5. Conozca los posibles signos y síntomas de la gripe aviar en seres humanos que han sido infectados por el virus H5N1 altamente patógeno:

   ▣ Fiebre

   ▣ Tos

   ▣ Dificultad para respirar

   ▣ Dolor de garganta

   ▣ Dolores musculares

   ▣ Conjuntivitis (dolor, inflamación, o enrojecimiento de ojos y párpados)

   ▣ Diarrea

6. Piense en ponerse la vacuna contra la gripe de la actual temporada.

7. Enseñe a los trabajadores a seguir todas las instrucciones y recomendaciones presentadas en esta Alerta.

Tome las siguientes medidas DURANTE un brote de gripe aviar:

1. Siga el plan de respuesta ante la gripe aviar.

2. Pregúntele a su médico si debe tomar medicamentos antivirales.

3. Utilice ropa protectora.

   ▣ Utilice las siguientes prendas protectoras si hay probabilidad de exposición a un virus de la gripe aviar:

      — Prendas exteriores (delantales y overoles)

      — Guantes

      — Protección para los pies (botas o fundas para botas)

      — Protección para la cabeza (sombreros o pañoletas)

(más al reverso)

■ Utilice prendas protectoras ligeras, desechables e impermeables.

■ Utilice guantes desechables, ligeros y de material impermeable (p. ej., de vinilo o de nitrilo de 8 a 12 mils) o guantes de goma gruesos de 18 mils, que se puedan volver a usar después de una desinfección.

■ Utilice fundas de botas desechables o botas que se puedan volver a usar después de desinfectarlas.

■ Utilice sombreros o pañoletas desechables para cubrir la cabeza.

4. Utilice equipo protector de ojos.

■ Cuando trabaje con aves de corral, utilice gafas de seguridad que no tengan orificios de ventilación o que tengan ventilación indirecta, un respirador con una máscara de cara completa, o un respirador con purificador de aire motorizado (PAPR) que tenga una capucha holgada o casco y visera protectora.

■ Retírese el equipo protector de los ojos con mucha precaución para evitar que las partes contaminadas hagan contacto con los ojos, la nariz o la boca.

5. Utilice respiradores.

■ Utilice un respirador de aire purificado con filtro para partículas, aprobado por NIOSH, cuando trabaje en granjas avícolas o si hay una exposición potencial a aves de corral infectadas o materiales y ambientes contaminados por el virus.

■ Utilice un filtro de partículas de una graduación igual o mayor a N–95.

6. Siga las instrucciones de un programa de protección respiratoria. Para asegurarse que los trabajadores cuentan con respiradores adecuados para protegerse de la gripe aviar, siga estas recomendaciones:

■ Designe a una persona calificada que supervise el programa y responda a las preguntas de los trabajadores.

■ Proporcione capacitación a los trabajadores en el uso del respirador y haga que se lo prueben, para asegurar que el respirador que cubre la cara se ajusta de manera cómoda y segura.

■ Incluya a todos los trabajadores que pueden estar en riesgo de exposición al virus de la gripe aviar.

7. Protéjase de una posible contaminación cuando se quite las prendas y el equipo protector.

■ Comience por quitarse la ropa protectora, sin retirarse el respirador, las gafas y los guantes.

— Coloque las prendas desechables en un recipiente[†] de seguridad para depositar estos desechos.

— Coloque las prendas reutilizables en recipientes[‡] con etiquetas o códigos por color para la limpieza y la desinfección.

■ Quítese los guantes y deposítelos en un recipiente de seguridad.

— Lávese otra vez minuciosamente las manos con agua y jabón.

— Si no hay instalaciones accesibles para lavarse las manos, utilice lociones o

---

[†]NIOSH recomienda el uso de envases seguros como aquellos descritos en el *Código de Regulaciones Federales* (Code of Federal Regulations) [29 CFR 1910.1030(d)(4)(iii)(B)]. Dichos envases por lo tanto deberán (1) poder cerrarse bien, (2) estar diseñados para guardar todo tipo de materiales y evitar la fuga de líquido durante su manejo, almacenamiento, transporte o envío, (3) estar identificados con etiquetas o códigos de color y (4) ser cerrados antes de su traslado.

[‡]NIOSH recomienda el uso de bolsas con etiquetas o códigos de color como aquellas descritas en el *Código de Regulaciones Federales* [29 CFR 1910.1030(d)(4)(iv)(A)(2)].

desinfectantes a base de alcohol provistos por su empleador.

— No utilice de manera excesiva las lociones o desinfectantes para las manos a base de alcohol, ya que pueden resecar la piel.

■ A continuación, retírese cuidadosamente las gafas y al final el respirador.

— Lávese minuciosamente las manos con agua y jabón.

— Si no hay instalaciones accesibles para lavarse las manos, utilice lociones o desinfectantes a base de alcohol provistos por su empleador.

8. Practique una buena higiene de las manos (uso de guantes, lavado de las manos con agua y jabón, o uso de desinfectantes para las manos) y aplique los procedimientos para la descontaminación que se han descrito arriba a fin de prevenir infecciones, evitar llevarse los virus a casa, y no permitir su propagación a otras granjas.

9. Dúchese al final de su jornada laboral y deje en el trabajo todas las prendas y equipo contaminado.

■ Dúchese en el lugar de trabajo o en una instalación cercana designada para la descontaminación.

■ Nunca utilice ropa o equipo protector contaminado afuera de su área de trabajo.

10. Participe en programas de monitorización y vigilancia de la salud.

■ Asegúrese de haber establecido un programa de vigilancia para identificar a los trabajadores sintomáticos en los 10 días siguientes a la exposición a aves infectadas o a materiales y ambientes contaminados por el virus.

■ Solicite atención médica inmediata para los trabajadores que presenten alguno de los síntomas siguientes en los 10 días posteriores a la exposición a aves infectadas o expuestas, o a materiales y ambientes contaminados por el virus:

— Fiebre

— Tos

— Dificultad para respirar

— Dolor de garganta

— Dolores musculares

— Conjuntivitis (infecciones de los ojos)

— Diarrea

■ Notifique al proveedor de atención médica que hubo una posible exposición a la gripe aviar antes de que la persona vaya a la consulta.

■ Reporte inmediatamente a los supervisores y las autoridades locales de salud los casos sospechosos de gripe aviar.

Para más información, sírvase leer la **Alerta de NIOSH: Proteja de la gripe aviar a los trabajadores de las granjas avícolas** [Publicación No. 2008–128 de DHHS (NIOSH) (reemplaza 2008–113)]. Obtenga copias gratuitas de esta Alerta a través de

1–800–CDC–INFO (1–800–232–4636)
TTY: 1–888–232–6348
Correo electrónico: cdcinfo@cdc.gov

o visite la página Web de NIOSH:
www.cdc.gov/spanish/niosh

Para recibir boletines mensuales actualizados de NIOSH visite el sitio Web **www.cdc.gov/niosh/eNews** y suscríbase a *NIOSH eNews*.

DEPARTAMENTO DE SALUD Y SERVICIOS HUMANOS
Centros para el Control y la Prevención de Enfermedades
Instituto Nacional para la Seguridad y Salud Ocupacional

Protecting Poultry Workers from Avian Influenza (Bird Flu)

> ### WARNING!
>
> **During an outbreak of avian influenza (bird flu), poultry workers* may become seriously ill or die after contact with infected poultry or contaminated materials.**

The National Institute for Occupational Safety and Health (NIOSH) requests help in protecting poultry workers from infection with viruses that cause avian influenza (also known as bird flu). Although human infection with avian influenza viruses is rare, workers infected with certain types of these viruses may become ill or die.

Some types of avian influenza viruses can cause serious illness or death in poultry and other birds. These viruses are referred to as *highly pathogenic viruses.* Rarely, these viruses may be passed to humans who contact infected poultry or virus-contaminated materials or environments.

The following workers are at risk of infection with highly pathogenic avian influenza viruses:

- Poultry growers and their workers

- Service technicians of poultry growers

- Workers at egg production facilities (caretakers, layer barn workers, and chick movers)

- Veterinarians and their staff who work with poultry

- Disease control and eradication workers on poultry farms (State, Federal, contract, and poultry farm workers)

This Alert describes the following:

1. Avian influenza in humans

2. Avian influenza outbreaks in chickens

3. Recommendations for protecting poultry workers from avian influenza

Remember these facts:

- No avian influenza epidemic now exists in humans.

- Scientists are currently most concerned about the highly pathogenic

---

*Poultry workers include all workers who may contact poultry or materials or environments contaminated by poultry.

avian influenza A virus known as H5N1. As of April 8, 2008, all human cases of influenza caused by this virus have occurred outside the United States.[†]

■ **The avian influenza virus rarely infects humans.**

■ **The avian influenza virus does not pass easily from person to person.**

NIOSH requests that owners and operators of poultry operations follow the recommendations in this Alert and use the controls presented here.

NIOSH also requests that safety and health officials, editors of trade journals, labor unions, and employers bring the recommendations in this Alert to the attention of all poultry workers and poultry farm operators.

BACKGROUND

What is avian influenza?

Avian influenza is caused by influenza A viruses and occurs worldwide in many species of birds. For this reason, avian influenza is often referred to as "bird flu."

Avian influenza viruses normally reside in the intestinal tracts (guts) of water and shore birds, and they usually cause little if any disease. Only a few of these viruses can cause disease in other animals and in humans—for example, the highly pathogenic H5N1 and H7N7 viruses.

Influenza A viruses are divided into two groups based on their pathogenicity (capacity to cause disease) to poultry:

■ *Low-pathogenic avian influenza viruses,* which cause a mild disease or no noticeable signs of disease

■ *Highly pathogenic avian influenza viruses,* which cause serious disease and high rates of death

Scientists are currently most concerned about the highly pathogenic avian influenza A virus known as H5N1.[‡] Although the H5N1 virus causes serious illness in birds, it has rarely infected humans.

The first known cases of human infection with H5N1 avian influenza occurred in 1997 in Hong Kong. Outbreaks of H5N1 avian influenza in poultry and some cases in humans began again in Asia in late 2003 and continue to be reported there. In addition, outbreaks in birds and humans have been reported in Africa, and outbreaks in birds have occurred in Europe. As of April 8, 2008, no outbreaks in birds or humans have been reported in North, Central, or South America.

Avian influenza outbreaks from the H5N1 virus have occurred in birds in more than 50 countries and in humans in 12 countries (see WHO [2008] at www.who.int/csr/disease/avian_influenza/en/).

Important avian influenza facts

Remember the following facts about the highly pathogenic H5N1 virus:

[†]For current information about outbreaks of avian influenza around the world, see www.cdc.gov/flu/avian/outbreaks/current.htm.

[‡]In this document, H5N1 always refers to the highly pathogenic form of avian influenza virus.

- Most human infections with this virus have resulted from contact with infected poultry or virus-contaminated materials or environments.

- H5N1 virus rarely infects humans.

- When this virus does infect humans, it does not pass easily from person to person—that is, transmission between humans has not been efficient or sustained [DHHS 2006].

- As of April 8, 2008, the H5N1 virus has not been detected in the United States.

- H5N1 virus can be spread from one location to another through

  — migrating birds (which may not show symptoms of disease) and

  — legal and illegal trade in poultry and other birds as well as their products.

- If the virus changes so that it can be easily passed from one person to another, it could cause a pandemic (worldwide) influenza outbreak in humans.

FREQUENTLY ASKED QUESTIONS

1. How are poultry affected by avian influenza?

Domestic poultry may be infected with either low-pathogenic or highly pathogenic viruses through contact with infected poultry, wild birds, or virus-contaminated materials or environments:

- Highly pathogenic avian influenza viruses spread quickly and may kill an entire poultry flock in 48 hours.

- Low-pathogenic avian influenza viruses may go unnoticed or cause only mild symptoms (such as ruffled feathers or a drop in egg production).

2. What is the risk of infection to humans?

Avian influenza viruses do not usually infect humans. However, 379 human cases of avian influenza A (H5N1) were reported to the World Health Organization (WHO) between late 2003 and April 8, 2008 [WHO 2008]. About 63% of these cases (239) were fatal. No human cases have been reported within North, Central, or South America.

H5N1 virus can be transmitted to people who contact infected poultry or virus-contaminated materials or environments. This type of transmission has not been frequent or sustained from one human to another.

Health risks related to human exposure to the low-pathogenic avian influenza viruses are poorly understood, but they are thought to be minimal. Only rare cases of human infection with low-pathogenic viruses have been reported. Nonetheless, anyone likely to have prolonged exposure to *any* avian influenza virus should take protective measures.

Photograph courtesy of U.S. Department of Agriculture.

Examples of workers at risk include the following:

- Poultry growers and their workers
- Service technicians of poultry growers
- Caretakers, layer barn workers, and chick workers at egg production facilities
- Veterinarians and their staff who work with poultry
- Workers involved in disease control and eradication on poultry farms (Federal, State, contract, and poultry farm workers)

Photograph copyright © Getty Images.

3. How is the virus passed to humans?

Avian influenza virus is excreted in the droppings, saliva, and nasal secretions of infected birds. The virus is believed to enter humans through the mouth, nose, or eyes. Scientists believe that the virus is most often passed to humans from contact with infected poultry that was sick or dead. Contact with the following materials or equipment may also be a source of infection:

- Droppings
- Feathers
- Litter
- Egg flats
- Cages

For more information about human infection with avian influenza viruses, see www.cdc.gov/flu/avian/gen-info/avian-flu-humans.htm.

4. Why are scientists concerned about the H5N1 virus?

Scientists are concerned about the H5N1 virus for the following reasons:

- H5N1 virus causes serious illness and death in poultry and therefore threatens domestic poultry throughout the world.
- This virus can cause serious illness and death in humans.
- If a strain of H5N1 changes so that it is highly infectious to humans and spreads easily from person to person, it could cause an influenza pandemic.

Public health authorities are monitoring outbreaks of human illness linked with avian influenza. To date, human infections with highly pathogenic avian influenza viruses identified since 1997 have not resulted in continued transmission from one person to another.

REPORTED OUTBREAKS

Current news about avian influenza deals mostly with human illness caused by the H5N1 virus. However, human infections have also been caused by other subtypes of avian influenza virus such as H7N7 and H7N3. The following reports describe outbreaks involving several subtypes of highly pathogenic avian influenza virus. One report describes an outbreak in poultry alone, with no reported human cases.

Report 1—Eighteen H5N1 human cases in Hong Kong, 1997

An outbreak of H5N1 avian influenza occurred in humans and poultry in Hong Kong during 1997. This outbreak involved 18 confirmed human cases, including six deaths [Chan 2002].

The first human case occurred in May, soon after outbreaks in poultry were reported at three farms. Seventeen more human cases occurred in November and December after infected poultry were found in wholesale and retail markets. Many of the infected humans had visited either a retail poultry stall or a live poultry market before becoming ill [Mounts et al. 1999]. All chickens and other poultry in Hong Kong were culled (destroyed) to stop the outbreak. No additional human cases were detected during this outbreak after the culling operation was complete.

Commercial poultry cullers and workers were not included among the 18 cases described here. However, laboratory tests showed that about 3% of poultry cullers and 10% of poultry workers showed evidence of earlier infection with H5N1 virus [Bridges et al. 2002].

Report 2—Eighty-nine H7N7 human cases in the Netherlands, 2003

In February 2003, a large outbreak of avian influenza was caused by the highly pathogenic H7N7 virus in commercial poultry farms in the Netherlands [Koopmans et al. 2004]. The infection spread to approximately 255 farms and resulted in the culling of all infected flocks (about 30 million chickens). The virus may have been introduced to the commercial flocks by infected wild ducks.

At the time of the outbreak, local authorities believed the risk to humans was low. However, 89 human infections were identified, with health complaints primarily consisting of conjunctivitis. Mild, influenza-like illness was associated with the conjunctivitis in a few cases.

However, one human fatality occurred in a veterinarian who had not received antiviral medication but had spent a few hours screening flocks that were later confirmed to be infected with the H7N7 virus. The highest risk of infection was in veterinarians and workers who culled infected poultry.

The outbreak was brought under control in about 2 months by culling infected flocks. An outbreak-management response team advised all workers who screened and culled poultry to wear goggles and respirators to reduce their exposures to the avian influenza virus. The team recommended that vaccination with the current flu vaccine be made mandatory for all poultry farmers and their families within a 3-kilometer radius of infected farms. They stressed the importance of hand washing and personal hygiene at home. Immediate treatment with oseltamivir (Tamiflu®) was recommended for all new conjunctivitis cases and a preventive dose (75 mg daily) was started for all persons handling potentially infected poultry. This dose was continued for 2 days after the last exposure.

Report 3—Two H7N3 human cases in Canada, 2004

On February 19, 2004, the Canadian Food Inspection Agency announced an outbreak of avian influenza in poultry from highly pathogenic H7N3 virus in the Fraser Valley region of British Columbia [Tweed et al. 2004; CDC 2006b]. Health Canada reported two cases of laboratory-confirmed H7N3 infections in humans. Both patients were poultry workers; one was involved in culling operations on March 13−14, 2004, and the other had close contact with poultry on March 22−23, 2004. Both patients developed conjunctivitis and other flu-like symptoms. Their illnesses resolved after treatment with antiviral medication (oseltamivir). Ten other poultry workers developed conjunctivitis symptoms and/or upper respiratory symptoms after contacting poultry. However, these infections were not laboratory-confirmed as H7N3 infections.

Culling operations by Federal workers and other measures were undertaken to control the spread of the virus. Authorities required personal protective equipment for all persons involved in culling activities. This equipment included N−95 respirators, gloves, goggles, biosafety suits, and footwear. Authorities also monitored compliance with prescribed safety measures. Epidemiologic, laboratory, and clinical surveillance was done for signs of avian influenza in exposed persons. However, no person-to-person transmission was detected during this outbreak.

Report 4—H5N2 in poultry, Texas, 2004: no human cases

In February 2004, an outbreak of avian influenza from highly pathogenic H5N2 virus was detected in a flock of 7,000 chickens in south-central Texas [Lee et al. 2005]. The chickens at the affected farm were being sold to live-bird markets in Houston. Approximately 1,700 chickens had been sold to the live-bird markets about a week before the laboratory confirmed avian influenza in

the flock. The flock was culled on February 21, 2004. No human infections were reported.

Report 5—Eight H5N1 human cases in Indonesia, 2006

Poultry in Indonesia and other nearby countries have suffered continuing outbreaks of illness from the H5N1 virus in 2006 and 2007. This virus is considered to be entrenched in poultry throughout much of Indonesia. This widespread presence of the virus and local conditions have resulted in a substantial number of human cases (102 cases since 2005).

In June 2006, Indonesia became the focus of media attention when H5N1 was identified in an outbreak involving eight members of an extended family in northern Sumatra [Butler 2006]. No samples were taken from the first patient, a 37-year-old woman who became ill on April 24 and died on May 4. However, samples from seven other family members confirmed the presence of H5N1 virus. Investigators assumed that the first patient was also infected with H5N1 virus (which she is thought to have contracted from infected poultry). In all, seven of the eight infected family members died. A 25-year-old brother of the first patient survived.

The outbreak was considered to be controlled on June 12, 2006—3 weeks after the death of the last victim—with no new cases reported. This cluster of H5N1 cases is the first instance in which WHO reported that human-to-human transmission may have occurred. Concerns over the cluster of cases have eased since no other large clusters of human cases have been identified.

CONCLUSIONS

Outbreaks in Birds

In birds, outbreaks of the H5N1 virus continue to spread in Europe, Asia, and Africa. These outbreaks are on a scale that has not been seen before. Continued worldwide spread of this virus will place poultry and poultry workers at increased risk of infection.

Human cases

Since January 2003, WHO has published the numbers of confirmed human illnesses and deaths from the H5N1 virus. Between January 2003 and April 8, 2008, WHO reported 379 confirmed human cases of infection with H5N1 virus in 14 countries—Azerbaijan, Cambodia, China, Djibouti, Egypt, Indonesia, Iraq, Laos, Myanmar, Nigeria, Pakistan, Thailand, Turkey, and Vietnam [WHO 2008]. Of these cases, 239 (63%) were fatal.

In 2007, Indonesia reported 42 new human cases of avian influenza, followed by Egypt (25), Vietnam (8), China (5), Laos (2), Cambodia (1), Myanmar (1), Nigeria (1), and Pakistan (1) [WHO 2008].

Human cases of avian influenza have most often been linked to close human contact with sick or dying poultry from backyard operations. Such contact is common in countries where poultry are numerous and birds are not generally confined by enclosures.

Continued sporadic infections of humans with H5N1 could increase the chances that the virus will change so that it can pass more easily from human to human. This change could result in an influenza pandemic.

Preventive steps

Additional efforts are needed to prevent new cases of avian influenza in humans. In Thailand, public health education campaigns and media reports about avian influenza have reached rural people at greatest risk [Olsen et al. 2005]. However, this information has not resulted in changed behavior to control risks for many Thai people. Culling flocks of ill birds has been highly effective in controlling some avian influenza outbreaks. But this preventive measure may not be effective in areas of Southeast Asia, where backyard flocks are common and poultry movement is difficult to control [CDC 2004; Olsen et al. 2005].

Poultry producers in the United States and around the world should take preventive steps to protect their workers and poultry flocks. Poultry producers can substantially reduce the risk to workers if they follow the recommendations listed in the following section.

RECOMMENDATIONS FOR PROTECTING POULTRY WORKERS

NIOSH recommends the following preventive steps for protecting poultry workers who are at risk of exposure to avian influenza viruses. These recommendations are discussed in more detail in the following subsections. Recommendations are intended for both poultry producers (owners and operators of poultry farms) and poultry workers.

Summary of recommendations

Take the following steps BEFORE an outbreak of avian influenza:

1. Make sure that an avian influenza response plan has been developed.

2. Follow the biosecurity practices presented in this Alert.

3. Know the signs of avian influenza in poultry.

4. Report sick or dying birds immediately.

5. Know the possible signs and symptoms of avian influenza in humans.

6. Consider getting the current season's flu shot.

7. Train workers in all guidance and recommendations presented in this Alert.

Take the following steps DURING an outbreak of avian influenza:

1. Follow the avian influenza response plan.

2. Ask your doctor about taking antiviral medication.

3. Wear personal protective clothing:
   — Outer garments (impermeable aprons or coveralls)
   — Gloves
   — Footwear (boots or boot covers)
   — Disposable head cover or hair cover

4. Wear eye protection (goggles or a full-facepiece respirator).

5. Wear a NIOSH-certified, air-purifying respirator with a particulate filter (N–95 or better).

6. Follow a written respiratory protection program.

7. Protect yourself when removing personal protective equipment and clothing.

8. Use good hand hygiene (proper use of gloves, hand-washing, and waterless hand sanitizers) and decontamination procedures.

9. Shower at the end of the work shift and leave all contaminated equipment and clothing at work.

10. Participate in health surveillance and monitoring programs.

Detailed recommendations

Take the following steps BEFORE an outbreak of avian influenza:

1. **Make sure that an avian influenza response plan has been developed to complement regional, State, and industry plans.**

   ▪ Use the following guidance to develop an avian influenza response plan:

      — CDC guidance in this Alert and at the following Web site: www.cdc.gov/flu/avian

      — The USDA national plan for responding to an outbreak of highly pathogenic avian influenza in the

Photograph copyright © Getty Images.

United States [APHIS 2007b]: www.aphis.usda.gov/newsroom/hot_issues/avian_influenza/avian_influenza_summary.shtml. This plan is intended to complement regional, State, and industry plans.

- ▨ Select a response plan manager.

- ▨ Coordinate your avian influenza response plan with appropriate State animal and public health agencies.

- ▨ Make sure that workers are aware of the avian influenza response plan and their responsibilities.

2. **Follow biosecurity practices to keep avian influenza and other diseases out of your poultry flock [APHIS 2007a]:**

- ▨ Keep your poultry flock isolated from outside environments.

- ▨ Prevent flocks from contacting wild birds and keep them away from water sources that might be contaminated by wild birds.

- ▨ Allow only essential workers and vehicles to enter the farm and poultry houses.

- ▨ Provide clean protective clothing, equipment, and disinfection facilities for workers.

- ▨ Thoroughly clean and disinfect equipment and vehicles entering and leaving the farm. Include tires and undercarriage.

  - — Use detergents and disinfectants (avian influenza viruses are sensitive to most).

  - — Use EPA-registered disinfectants that are labeled as effective against influenza viruses.

  - — Use heating and drying (which inactivate the viruses).

- ▨ Do not borrow equipment or vehicles from other farms and do not lend yours.

- ▨ Avoid visiting other poultry farms.

- ▨ If you do visit another farm or a live-bird market, change footwear and clothing before contacting your own flock again.

- ▨ Do not bring birds from slaughter channels (especially live-bird markets) back to the farm.

3. **Know the signs of avian influenza in poultry.**

Be aware of the signs of avian influenza infection in poultry so that you can do the following:

- ▨ Recognize sick birds

- ▨ Quarantine the farm to prevent the spread of disease

- ▨ Protect workers from infection

In domestic poultry, signs of infection with the highly pathogenic H5N1 virus may vary depending on the viral strain, age and species of bird, other existing diseases in the poultry, and environment. The signs may include the following:

- ▨ Sudden death without clinical signs or symptoms

- ▨ Lack of coordination

- ▨ Lack of energy and appetite

- Soft-shelled or misshapen eggs

- Decreased egg production

- Purple discoloration of the wattles, combs, and legs

- Swelling of the head, eyelids, combs, wattles, and hocks

- Diarrhea

- Nasal discharge

- Coughing and sneezing

Some birds may be infected with avian influenza but appear to be healthy.

4. **Report sick or dying birds immediately.**

  - Immediately report any sick or dying birds in your poultry flock to the proper authorities:

    — Call the U.S. Department of Agriculture (USDA) Veterinary Services toll free at

      1–866–536–7593

    — Or contact your State veterinarian or local extension agent.

  - **Don't wait to report sick birds.** Early detection of avian influenza is essential to prevent its spread.

5. **Know the possible signs and symptoms of avian influenza in humans.**

  Know the signs and symptoms of avian influenza in humans infected with the highly pathogenic H5N1 virus so that ill persons can be treated immediately:

  - Fever

  - Cough

- Shortness of breath

- Sore throat

- Muscle aches

- Conjunctivitis (redness, swelling, and pain in the eyes and eyelids)

- Diarrhea

Watch for these signs and symptoms of avian influenza for up to 10 days after exposure to infected or exposed birds or to virus-contaminated materials or environments.

So far, conjunctivitis has been extremely rare in humans infected with the highly pathogenic H5N1 virus—but it is a common symptom in humans infected with the highly pathogenic H7N7 virus. Avian influenza can also lead to pneumonia, acute respiratory distress, and other life-threatening complications.

6. **Consider getting the current season's flu shot.**

CDC recommends the current season's flu shot for workers involved in avian influenza control activities. Other poultry workers should also consider getting the current flu shot. Although a flu shot will not prevent infection with avian influenza, it *could* prevent dual infection—that is, infection with both an avian influenza virus *and* a human influenza virus at the same time. Such dual infection might result in the formation of new viral strains. If one of these new strains passes easily from person to person, an influenza pandemic could result.

For information about dual infection, use of antiviral medications, and vaccination of poultry workers, see the CDC Web site on avian influenza at www.cdc.gov/flu/avian/index.htm.

7. **Train workers in all guidance and recommendations presented in this Alert.**

Take the following steps DURING an outbreak of avian influenza:

1. **Follow the avian influenza response plan.**

2. **Ask your doctor about taking antiviral medication.**

Before you begin disease control activities during an outbreak of avian influenza, ask your doctor about taking antiviral medication. The Centers for Disease Control and Prevention (CDC) recommends that workers receive an influenza antiviral drug daily for the entire time they are in direct contact with infected poultry or with virus-contaminated materials or environments [CDC 2006a]. In addition, the Occupational Safety and Health Administration (OSHA) recommends that workers take the antiviral drug for 1 week following exposure [OSHA 2006].

Oseltamivir is currently the antiviral drug most often used for influenza. This drug is preferred because the avian influenza virus is less likely to be resistant to it than to amantadine or rimantadine (two other drugs used to prevent or treat influenza A). A fourth drug, zanamivir, may be considered as an alternative to oseltamivir for prophylaxis when available [Hayden and Pavia 2006].

3. **Wear personal protective clothing.**

Personal protective clothing is clothing that protects the torso (aprons, outer garments, or coveralls), hands (gloves), feet (boots or boot covers), and head (head covers or hair covers) from exposure to harmful agents. Many poultry workers routinely wear personal protective clothing.

Poultry workers should be required to wear personal protective clothing whenever they may be exposed to avian influenza viruses. Such clothing will prevent skin contact with virus-contaminated materials or environments. It will also reduce the chances of carrying contaminated material outside the poultry barn or worksite.

Outer garments. When selecting protective outer garments such as aprons or coveralls, take the following steps:

■ Select impermeable, disposable protective clothing when possible.

■ Select lightweight clothing when appropriate to protect workers from heat stress. For example, choose a lightweight impermeable coverall instead of a chemical-resistant suit if possible.

Gloves. Gloves may be lightweight and disposable (8- to 12-mil nitrile or vinyl, for example), or they may be heavy duty rubber (18 mils thick or greater) and reusable after disinfection. Gloves should be waterproof. When selecting gloves, consider the following:

■ Activities performed by the worker

■ Dexterity requirements

■ Need for glove durability and resistance to tearing and abrasion

Regardless of the type of gloves selected, make sure they do not make existing dermatitis worse or damage healthy skin from prolonged exposure to water or sweat. Wearing a thin cotton glove under a protective outer glove may prevent dermatitis.

Foot protection. Select disposable boot covers or boots that can be disinfected. These will protect workers from contact with harmful agents and will prevent them from being carried from one location to another.

Head protection. Select disposable, lightweight head covers or hair covers.

Sources of personal protective clothing and equipment. For sources and manufacturers of personal protective clothing or other personal protective equipment, see the *Buyer's Guide* of the International Safety Equipment Association [www.safetyequipment.org].

4. **Wear eye protection.**

    Eye protection is important to prevent eye contact with virus-contaminated dusts, droplets, and aerosols and to keep workers from touching their eyes with contaminated fingers or gloves.

    ■ When working with poultry, wear unvented or indirectly vented safety goggles, a respirator with a full facepiece, or a powered, air-purifying respirator (PAPR) with a loose-fitting hood or helmet and face shield.

    ■ If you wear safety goggles, make sure they are either

       — *unvented* (eyecup goggles, for example) or

       — *indirectly vented.*

       If indirectly vented goggles are properly fitted and have a good antifog coating, they may be used by poultry workers with a low risk of exposure to avian influenza. However, such goggles are not airtight and will not prevent exposures to airborne material.

    ■ Do not use *directly vented* goggles or safety glasses for working with infected birds. They do not protect workers from fine particles, splashes, or aerosols.

    ■ If you need *prescription lenses*, use

       — protective eyewear with built-in prescription lenses,

       — lens inserts,

       — protective eyewear that fits snugly over prescription glasses without changing their position or obstructing vision (such as full-facepiece respirators, PAPRs with hoods or helmets, and some styles of goggles), or

       — contact lenses with goggles, a respirator with a full facepiece, or a PAPR with a loose-fitting hood or helmet and face shield.

    ■ Fit eye protection and respirators at the same time:

       — Some goggles can change the fit of a full-facepiece respirator.

       — Eye protection may interfere with the seal of a half-facepiece respirator.

    ■ Wear your eye protection or prescription glasses when you check the seal of a respirator before each use.

Glasses should not protrude into the seal area of a full-facepiece respirator.

■ Remove eye protection carefully to prevent contaminated equipment from contacting eyes, nose, or mouth.

For more information about eye safety, see www.cdc.gov/niosh/topics/eye.

5. **Wear a NIOSH-certified, air-purifying respirator with a particulate filter (N–95 or better).**

In agricultural environments, respirators are important to prevent exposures to viruses as well as to other agents such as bacteria, fungi, and endotoxins.

■ Wear a NIOSH-certified, air-purifying respirator with a particulate filter (N–95 or better) whenever you are working in poultry barns or may be exposed to infected poultry or virus-contaminated materials or environments. These are the most practical and appropriate respirators for such use.

■ See Table 1 to compare the costs and advantages of the five types of air-purifying respirators.

■ See *NIOSH Respirator Selection Logic 2004* [NIOSH 2005] and *Histoplasmosis—Protecting Workers at Risk* [NIOSH 2004] for more information about selecting and using respirators for infectious agents.

6. **Follow a written respiratory protection program.**

To make sure that respirators protect workers from avian influenza, do the following:

■ Designate a person trained in the selection, use, and fitting of respirators to oversee the program and answer workers' questions.

■ Provide workers with respirator training and fit testing to assure a safe and comfortable seal for the respirator facepiece.

■ Include all workers who may be at risk of exposure to avian influenza virus.

■ *See Safety and Health Topics: Respiratory Protection* [OSHA 2007] at www.osha.gov/SLTC/respiratoryprotection/index.html for more information about respiratory protection programs and respirators.

7. **Protect yourself when removing personal protective clothing and equipment.**

Protect yourself and prevent the avian influenza virus from spreading to other areas by taking these steps when removing protective clothing and equipment:

■ With your respirator, goggles, and gloves on, remove all personal protective clothing.

— Place disposable clothing in a secure container.§

— Place reusable clothing in labeled or color-coded bags or

---

§NIOSH recommends the use of secure containers such as those described in the *Code of Federal Regulations* [29 CFR 1910.1030(d)(4)(iii)(B)]. Such containers should therefore be (1) closable, (2) constructed to contain all contents and prevent leakage of fluids during handling, storage, transport, or shipping, (3) labeled or color-coded, and (4) closed before removal.

containers** for cleaning and disinfection.

- Remove gloves and discard them in a secure container.

  — Thoroughly wash your hands with soap and water.

  — If no hand-washing facilities are available, use waterless soaps or alcohol-based sanitizers provided by your employer.

- Next, carefully remove your goggles and then your respirator.

- Thoroughly wash your hands again with soap and water. If no hand-washing facilities are available, use waterless soaps or alcohol-based sanitizers provided by your employer.

8. **Use the good hand hygiene and decontamination procedures outlined here to prevent infection, avoid taking viruses home, and keep them from spreading to other farms:**

- Wear gloves whenever you may be exposed to infected poultry.

- Remove your gloves immediately after work and after removing protective clothing. Dispose of gloves in a secure container.

- Wash your hands thoroughly with soap and water.

- If no hand-washing facilities are available, use waterless soaps or alcohol-based sanitizers provided by your employer.

- Be careful about using waterless soaps or alcohol-based sanitizers too often. They can be very harsh to the skin.

9. **Shower at the end of the work shift and leave all contaminated clothing and equipment at work.**

- Shower at the worksite or at a nearby decontamination station.

- Never wear contaminated clothing or equipment outside the work area.

10. **Participate in health surveillance and monitoring programs.**

- Make sure a surveillance program has been established to identify workers who develop symptoms of avian influenza.

- Seek immediate medical care for workers who develop any of the following symptoms within 10 days of exposure to infected or exposed birds or to virus-contaminated materials or environments:

  — Fever

  — Cough

  — Shortness of breath

  — Sore throat

  — Muscle aches

  — Conjunctivitis (eye infections)

  — Diarrhea

- Tell the health care provider about the possible avian influenza exposure before the ill person arrives.

**NIOSH recommends the use of labeled or color-coded bags such as those described in the *Code of Federal Regulations* [29 CFR 1910.1030(d)(4)(iv)(A)(2)].

- Promptly report suspected human cases to supervisors and to local health authorities.

ACKNOWLEDGMENTS

The principle contributors to this Alert were Greg Kullman, Ph.D., C.I.H.; Lisa J. Delaney, M.S., C.I.H.; John Decker, M.S., C.I.H.; Kathleen MacMahon, M.S., D.V.M.; and Anne Hamilton. Gino Fazio and Vanessa Becks provided desktop design and production.

Please direct comments, questions, or requests for additional information to the following:

David Weissman, M.D.
Director, Division of Respiratory Disease
  Studies
National Institute for Occupational Safety
  and Health
1095 Willowdale Road
Morgantown, WV 26505-2888
Telephone: 304-285-5749

Or call 1-800-CDC-INFO (1-800-232-4636) (TTY: 1-888-232-6348)

We greatly appreciate your assistance in protecting the health of U.S. workers.

John Howard, M.D.
Director, National Institute for
  Occupational Safety and Health
Centers for Disease Control and
  Prevention

REFERENCES CITED

APHIS [2007a]. Biosecurity for the birds. Washington, DC: Animal and Plant Health Inspection Service, United States Department of Agriculture [www.aphis.usda.gov/vs/birdbiosecurity/hpai.html].

APHIS [2007b]. Draft summary of the national avian influenza (AI) response plan, August 2007. Washington, DC: Animal and Plant Health Inspection Service, U.S. Department of Agriculture [www.aphis.usda.gov/newsroom/hot_issues/avian_influenza/avian_influenza_summary.shtml].

Bridges CB, Lim W, Hu-Primmer J, Sims L, Fukuda K, Mak KH, Rowe T, Thompson WW, Conn L, Lu X, Cox NJ, Katz JM [2002]. Risk of influenza A (H5N1) infection among poultry workers, Hong Kong, 1997–1998. J Infect Dis *185*:1005–1010.

Butler D [2006]. Family tragedy spotlights flu mutations. Nature *44*(13): 114–115.

CDC (Centers for Disease Control and Prevention) [2004]. Cases of influenza A (H5N1)—Thailand, 2004. MMWR *53*:100–103.

CDC (Centers for Disease Control and Prevention) [2006a]. Interim guidance for protection of persons involved in U.S. avian influenza outbreak disease control and eradication activities [www.cdc.gov/flu/avian/professional/protect-guid.htm].

CDC (Centers for Disease Control and Prevention) [2006b]. Past avian influenza outbreaks [www.cdc.gov/flu/avian/outbreaks/past.htm#h7n3canada].

CFR. Code of Federal regulations. Washington, DC: U.S. Government Printing Office, Office of the Federal Register.

Chan PKS [2002]. Outbreak of avian influenza A (H5N1) virus infection in Hong Kong in 1997. Clin Infect Dis *34*(Suppl 2):S58–S64.

DHHS (U.S. Department of Health and Human Services) [2006]. Indonesia situation update—May 31 [www.pandemicflu.gov/news/indonesiaupdate.html].

Hayden F, Pavia A [2006]. Antiviral management of seasonal and pandemic influenza. J Infect Dis *194*(Suppl 2):S119–S126.

Koopmans M, Wilbrink B, Conyn M, Natrop G, van der Nat H, Vennema H, Meijer A, van Steenbergen J, Fouchier R, Osterhaus A, Bosman A [2004]. Transmission of H7N7 avian influenza A virus to human beings during a large outbreak in commercial poultry farms in the Netherlands. Lancet *363*:587–593.

Lee CW, Swayne DE, Linares JA, Senne DA, Suarez DL [2005]. H5N2 avian influenza outbreak in Texas in 2004: the first highly pathogenic strain in the United States in 20 years? J Virol *79*:11412–11421.

Mounts AW, Kwong H, Izurieta HS, Ho Y, Au T, Lee M, Buxton Bridges C, Williams SW, Mak KH, Katz JM, Thompson WW, Cox NJ, Fukuda K [1999]. Case-control study of risk factors for avian influenza A (H5N1) disease, Hong Kong, 1997. J Infect Dis *180*(2):505–508.

NIOSH [2004]. Histoplasmosis—protecting workers at risk. Cincinnati, OH: U.S. Department of Health and Human Services, Centers for Disease Control and Prevention, National Institute for Occupational Safety and Health,

DHHS (NIOSH) Publication No. 2005–109 [www.cdc.gov/niosh/docs/2005-109].

NIOSH [2005]. NIOSH respirator selection logic 2004. Cincinnati, OH: U.S. Department of Health and Human Services, Centers for Disease Control and Prevention, National Institute for Occupational Safety and Health, DHHS (NIOSH) Publication No. 2005–100 [www.cdc.gov/niosh/docs/2005-100/default.html].

Olsen SJ, Laosiritaworn Y, Pattanasin S, Prapasiri P, Dowell SF [2005]. Poultry-handling practices during avian influenza outbreak, Thailand. Emerg Infect Dis *11*:1601–1603.

OSHA [2006]. Protecting employees from avian flu (avian influenza) viruses. Washington, DC: U.S. Department of Labor, Occupational Safety and Health Administration [www.osha.gov/OshDoc/data_AvianFlu/avian_flu_guidance_english.pdf].

OSHA [2007]. Safety and health topics: respiratory protection. Washington, DC: U.S. Department of Labor, Occupational Safety and Health Administration [www.osha.gov/SLTC/respiratoryprotection/index.html].

Tweed SA, Skowronski DM, David ST, Larder A, Petric M, Lees W, Li Y, Katz J, Krajden M, Tellier R, Halpert C, Hirst M, Astell C, Lawrence D, and Mak A [2004]. Human illness from avian influenza H7N3, British Columbia. Emerg Infect Dis *10*:2196–2199.

WHO [2008]. Cumulative number of confirmed human cases of avian influenza A/(H5N1) Reported to WHO [www.who.int/csr/disease/avian_influenza/country/en/].

Appendix A

Advantages, Disadvantages, and Costs of Air-purifying Respirators for Protecting Poultry Workers[*]

Respirator type[†] and APF[‡]	Advantages	Disadvantages	Cost (2004 dollars)
Filtering-facepiece respirator (disposable; dust mask); APF = 10	▪ Is lightweight. ▪ Needs no maintenance or cleaning. ▪ Has no effect on mobility.	▪ Provides no eye protection. ▪ Provides no protection against irritant gases such as ammonia. ▪ Can add to heat burden. ▪ Permits inward leakage at gaps in face seal. ▪ Does not have adjustable head straps on many models. ▪ Is difficult for a user to do a seal check. ▪ Varies greatly in level of protection provided by different models. ▪ May make communication difficult. ▪ Requires fit testing to select proper facepiece size. ▪ May not fit properly when used with some eyewear.	$0.70 to $10
Elastomeric half-facepiece respirator; APF = 10	▪ Requires low maintenance. ▪ Has reusable facepieces and replaceable filters and cartridges. ▪ Permits use of dual cartridges to protect workers from exposures to particles, gases, and vapors. ▪ Has no effect on mobility.	▪ Provides no eye protection. ▪ Can add to heat burden. ▪ Permits inward leakage at gaps in face seal. ▪ Requires cleaning and disinfection of facepiece before reuse and thus poses a contact exposure risk. ▪ May make communication difficult. ▪ Requires fit testing to select proper facepiece size. ▪ May not fit properly when used with some eyewear.	Facepiece: $12 to $35 Filters: $4 to $8 each

(Continued)

See footnotes at end of table.

(Continued). Advantages, Disadvantages, and Costs of Air-purifying Respirators for Protecting Poultry Workers[*]

Respirator type[†] and APF[‡]	Advantages	Disadvantages	Cost (2004 dollars)
Powered, air-purifying respirator (PAPR) with hood, helmet, or loose-fitting facepiece; APF = 25	▪ Provides eye protection. ▪ Provides protection for people with beards, missing dentures, or facial scars. ▪ Has low breathing resistance. ▪ Has combination cartridges that can be used for exposures to particles, gases, and vapors. ▪ Creates a cooling effect with flowing air. ▪ Has face seal leakage that is generally outward. ▪ Requires no fit testing. ▪ Permits wearing of prescription glasses. ▪ Permits better communication than rubber half-facepiece or full-facepiece respirators. ▪ Has reusable components and replaceable filters.	▪ Has added weight from battery and blower. ▪ Is awkward to wear for some tasks. ▪ Requires cleaning and disinfection of components before reuse and thus poses a contact exposure risk. ▪ Requires battery charging. ▪ Requires air-flow testing with flow device before use.	Unit: $400 to $1,000 Filters: $10 to $30
Elastomeric, full-facepiece respirator with N–100, R–100, or P–100 filters; APF = 50	▪ Provides eye protection. ▪ Requires low maintenance. ▪ Has reusable facepieces and replaceable filters and cartridges.	▪ Can add to heat burden. ▪ Has reduced field of vision compared with a half-facepiece respirator. ▪ Permits inward leakage at gaps in face seal.	Facepiece: $90 to $240 Filters: $4 to $8 Each nose cup: $30

(Continued)

See footnotes at end of table.

(Continued). Advantages, Disadvantages, and Costs of Air-purifying Respirators for Protecting Poultry Workers*

Respirator type[†] and APF[‡]	Advantages	Disadvantages	Cost (2004 dollars)
	▪ Has combination cartridges that can be used for exposures to particles, gases, and vapors. ▪ Has no effect on mobility. ▪ Has a more effective face seal than a filtering-facepiece or rubber half-facepiece respirator.	▪ Requires cleaning and disinfection of facepiece before reuse and thus poses a contact exposure risk. ▪ Requires fit testing to select proper facepiece size. ▪ May require nose cup or lens treatment to prevent fogging of facepiece lens. ▪ Requires spectacle kit for users who wear prescription glasses.	
Powered, air-purifying respirator (PAPR) with tight-fitting half facepiece or full facepiece; APF = 50	▪ Provides eye protection with full facepiece. ▪ Has low breathing resistance. ▪ Has face seal leakage that is generally outward. ▪ Creates a cooling effect with flowing air. ▪ Has reusable components and replaceable filters. ▪ Has combination cartridges that can be used for exposures to particles, gases, and vapors.	▪ Has added weight from battery and blower. ▪ Is awkward to wear for some tasks. ▪ Provides no eye protection with a half facepiece. ▪ Requires cleaning and disinfection of components before reuse and thus poses a contact exposure risk. ▪ Requires fit testing to select proper facepiece size. ▪ Requires charging of battery. ▪ May make communication difficult. ▪ Requires spectacle kit for people who wear prescription glasses with full-facepiece respirators. ▪ Requires air-flow testing with flow device before use.	Unit: $500 to $1,000 Filters: $10 to $30

*All respirator types mentioned here meet the minimum requirements for N-95 respirators.
[†]Alternative filter types may be obtained for each type of respirator described here.
[‡]APF = assigned protection factor.

NOTES

NOTES